THE
POWER
OF

How Machine Learning is Changing
the World as We Know It

MARK HRVATIN

**Copyright © 2023 Mark Hrvatin.
All rights reserved.**

Reproducing, duplicating, or otherwise transmitting any of the information included in this book is strictly prohibited unless express written permission is first obtained from the author or the publisher.

In no event will the publisher or the author be held liable in any way, shape, or form, whether directly or indirectly, for any damages, reparations, or financial losses that may have been incurred as a direct or indirect result of the material included within this book.

Legal Notice

Copyright regulations guard this book. It can only be used for one's purposes. You are not permitted to alter, distribute, sell, use, the quote from, or paraphrase any portion of this book or the material included within it without first obtaining permission from either the author or the publisher.

Disclaimer Notice

Please take into consideration that the material presented in this text is only intended for the sake of edification and amusement. Information that is accurate, up-to-date, dependable, and comprehensive has been presented with every effort that may be made. However, there are no guarantees, either express or implied, of any sort. The reader understands and agrees that the author is not involved in the practice of delivering professional, legal, financial, or medical advice. This book contains information that was obtained from a wide variety of different sources. Therefore, before performing any of the methods described in this book, it is strongly recommended that you first get the advice of a qualified expert.

CONTENTS

INTRODUCTION..7

CHAPTER 1: UNDERSTANDING THE BASIC OF AI..11

    1.1 The Birth and Growth of Artificial Intelligence..12

    1.2: What can AI do for Human?..................13

    1.3: What is Machine Learning?..................15

    1.4: Practical Uses of Machine Learning in Real-life Scenarios..19

    1.5 What is Natural Language Processing (NLP)?..32

CHAPTER 2: COMPUTER VISION..................45

    2.1 Image Recognition..................................46

    2.2 Object Detection....................................49

    2.3 Image Segmentation.............................52

    2.4 Video Analysis.......................................55

CHAPTER 3: AI IN ACTION 59

3.1: AI in Healthcare   60

3.2: AI in Finance   62

3.3 AI in Transportation 63

3.4: AI in Retail   64

3.5: AI in Manufacturing   66

3.6 AI in Marketing   70

CHAPTER 4: THE ROLE OF AI IN ROBOTS 73

4.1 What is Robotics? 75

4.2 Are Robotics and AI the Same Thing? 76

4.3 What is The Role of AI in Robotics? 77

4.4 Some Examples of AI applied to Robotics 79

CHAPTER 5: AI AND SOCIETY 81

5.1 The Impact of AI on Employment 82

5.2 The Fusion of AI in Education 83

5.3 Revolution of Healthcare with AI 84

5.4 AI and Ethics 86

5.5 AI and the Environment 87

5.6 Bias and Fairness in AI 89

5.7 AI and Privacy 91

5.8 Legal Liability..................................92

CHAPTER 6: FUTURE OF AI..........................95
   6.1 Trends and Developments in AI................97
   6.2 Opportunities and Challenges in AI........98
   6.3 AI and the Singularity............................102
   6.4 Quantum Computer and AI....................103
   6.4 AI and the Singularity............................106

CONCLUSION..109

BIBLIOGRAPHY...113

INTRODUCTION

THE POWER OF AI

Artificial intelligence, or AI, describes a machine's capacity to carry out operations that ordinarily call for human intelligence, such as comprehending natural language, spotting patterns, and picking up knowledge through experience. AI systems can be taught to carry out certain activities, like playing chess or operating a vehicle, as well as educated to spot patterns and make predictions.

The concept of artificial intelligence has existed for millennia, but the contemporary area of AI is believed to have begun around the middle of the 20th century. In the 1950s and 1960s, researchers developed computer systems capable of simulating human thought processes, such as logical reasoning and problem solving.

McCarthy, who invented the term "artificial intelligence" in 1956 and helped design the first AI programming language, Lisp, was one of the pioneers of modern AI. Marvin Minsky, who co-founded the MIT AI Lab in 1959, and Allen Newell and Herbert Simon, who developed the General Problem Solver software in the 1950s, were also essential in the early development of artificial intelligence.

Since then, the field of artificial intelligence has exploded in size and scope, spawning several offshoots like machine learning, Natural language understanding, Computer vision, and robots. AI is now employed across all industries, from driverless cars and medical diagnostics to voice assistants like Siri and Alexa.

AI now plays a significant role in our contemporary society, influencing a variety of facets of our lives, including entertainment, banking, healthcare, and transportation. One of the key components of AI is Machine Learning (ML), A specialized sector of AI that prioritizes creating algorithms that can learn and improve from data without explicit programming. ML has enabled computers to perform complex tasks that were once only possible by humans, such as recognizing images, translating languages, and making predictions. The development of autonomous systems, including self-driving cars, drones, and robotics, has also benefited from machine learning.

Businesses are using AI and ML to personalise their offerings, streamline operations, and make data-driven decisions, which is also changing how firms function. Ultimately, AI and ML have boundless potential, and their influence on society will only increase over time.

This book examines the potential impact of AI on our current reality. We'll look at the fundamentals of machine learning and how it functions as well as actual applications of AI in various fields. We will also look at how AI will affect society as a whole, including how work will be done in the future and issues with ethics, privacy, and security.We

THE POWER OF AI

will demonstrate throughout this book how AI is more than just a buzzword; it is a revolutionary technology that is changing the way we live. This book will provide you the information and insights you need to grasp the power of AI and its potential to improve our world, whether you're an entrepreneur, a corporate leader, or simply interested in the future.

CHAPTER 1
UNDERSTANDING THE BASIC OF AI

1.1 The Birth and Growth of Artificial Intelligence

Artificial intelligence has been around since the middle of the twentieth century. The Dartmouth Conference, which was convened by John McCarthy, Marvin Minsky, and others in 1956, is usually regarded as the origin of artificial intelligence. Researchers at the time saw AI as a way to simulate human cognition, thus they focused on developing programs that could do tasks like logical deduction and problem solving.

In the decades that followed, AI research progressed in fits and starts. The 1960s saw a surge of interest in AI, with researchers developing new techniques like neural learning and machine networks. However, progress slowed in the 1970s and 80s, as the limitations of existing techniques became apparent.

AI research began to pick up again in the 1990s and 2000s, as advances in computer processing power and data storage enabled researchers to develop more sophisticated algorithms. AI is now used in a wide range of applications, from natural language and speech recognition processing to self-driving cars and medical diagnosis systems.

Notwithstanding the considerable strides made in AI in recent years, there is still much work to be done. Researchers are still debating fundamental issues such as how to build machines that can truly understand and reason about the world. As AI continues to evolve and improve, it is likely to have a profound impact on the way we live and work.

1.2: What can AI do for Human?

Artificial intelligence's potential extends to a wide range of applications, assisting humans in a variety of ways. One of the main advantages of AI is its ability to automate repetitive or mundane tasks, freeing up humans to focus on more creative and high-level work. For example, AI-powered chatbots can handle basic customer service inquiries, allowing human support agents to focus on more complex issues.

AI can also be used to improve decision-making by processing huge quantities of information and detecting patterns that humans may be unable to detect on their own. This can be especially beneficial in fields such as healthcare, where AI algorithms can assist doctors in diagnosing diseases and developing treatment plans based on patient data.

Another way AI can assist humans is to improve safety and security. Self-driving cars, for example, have the potential to greatly reduce the number of accidents caused by human error. AI-powered security systems can also aid in the prevention of

THE POWER OF AI

crime by detecting unusual behaviour and alerting authorities to potential threats.

AI can also help people with disabilities by providing tools such as speech recognition and text-to-speech technology. This can make it easier for people who are blind, deaf, or have mobility problems to navigate the world.

However, it is relevant to state that AI is not a magic solution to all our problems. There are still many challenges to overcome, such as ensuring that AI is used ethically and responsibly, and that it does not perpetuate biases or discriminate against certain groups. But, with adequate control and regulation, AI has the potential to be a potent tool for assisting humans in solving complicated issues and enhancing their lives.

> ### *Did you know?*
> AI learns like humans: Neural networks, a type of AI inspired by the human brain, allow AI systems to learn and adapt over time by processing information and adjusting connections, similar to the way humans acquire knowledge and skills.

1.3: What is Machine Learning?

Machine learning is a type of artificial intelligence that involves training computer algorithms to recognize patterns and make predictions or decisions based on data. It uses data driven strategies to allow the computer to learn and improve its performance over the period of time without being specifically programmed for each individual task. Essentially, it allows machines to enhance their proficiency in tasks like identifying objects in pictures, comprehending speech, or making predictions by learning from data. It has a wide range of uses in many industries, such as healthcare, finance, and engineering, and it keeps getting better and better as new methods and algorithms are created.

Machine learning encompasses various types, such as supervised learning, unsupervised learning, and reinforcement learning.

- **Supervised learning**

Supervised learning is a machine learning technique that involves the use of labelled data to train a model. In supervised learning, the data is labelled with the correct output or target variable, which the algorithm uses to learn patterns and relationships within the data. The goal of supervised learning is to predict the correct output for new input data that the model has not yet seen.

Supervised learning algorithms can be broadly classified into two categories: regression and classification. Regression models are used when the target variable is continuous, while classification models are used when the target variable is categorical.

The training process in supervised learning involves the use of an objective function, or loss function, which measures the difference between the predicted output and the actual output. Using an optimization algorithm like gradient descent, which modifies the model parameters to minimize the loss function, the objective function is optimized.

The potential practical applications of supervised learning algorithms are numerous, spanning fields such as image recognition, natural language processing, speech recognition, and predictive analytics in finance and healthcare. One of the main advantages of the supervised learning technique is that it allows for accurate predictions on new, unseen data. However, a disadvantage of supervised learning is that it necessitates a significant amount of labelled data to properly train the model.

- **Unsupervised learning**

Unsupervised learning is a technique that employs machine learning algorithms to analyze and cluster datasets that are not labeled. These algorithms are capable of finding latent patterns or groupings of data without any human involvement. Unsupervised learning is particularly useful for exploratory data analysis, customer segmentation, cross-selling techniques, and image recognition, owing to its capacity to identify similarities and differences in information.

For example, let's say unsupervised learning algorithm is presented with a dataset that consists of images of various types of cats and dogs. The algorithm has not been trained on this dataset, and therefore it has no prior knowledge of the characteristics of the dataset. The goal of the unsupervised learning algorithm is to identify the features of the images on its own. To accomplish this, the algorithm clusters the image dataset into groups based on similarities between images. By analyzing the similarities between images, the algorithm can identify key features of the cats and dogs in the dataset without any human involvement or labeled data.

- **Reinforcement learning**

Reinforcement Learning or RL is a category of machine learning that empowers an AI agent to learn from its experiences via trial and error. The agent receives feedback in the form of rewards or penalties for its actions, with the goal of optimizing the reward function. This method of education aims to closely resemble natural intelligence. It can be used in a wide range of industries.

While reinforcement learning shares some similarities with supervised learning, the main difference is that the feedback in reinforcement learning does not provide the appropriate series of actions that the agent needs to follow. Instead, the agent must determine its own actions to successfully complete the task. In contrast, unsupervised learning aims to identify similarities and differences (variations or contrasts) among data points, whereas the goal of reinforcement learning is to find the best action model that maximizes the cumulative reward for the agent. Unlike supervised learning, which relies on a training dataset, reinforcement learning uses the agent's interactions with the environment and the feedback it receives to solve the problem at hand.

- **Deep learning**

The use of artificial neural networks to model and solve complex problems It is called "deep" learning because it involves neural networks with multiple layers, which can allow for more complex representations of data and more accurate predictions. This technology is a vital component of data science, which also includes statistics and predictive modelling. Data scientists, who are responsible for gathering, analyzing, and interpreting large volumes of data, find deep learning to be an invaluable tool, as it accelerates and simplifies the process. In essence, deep learning automates predictive analytics. Unlike conventional machine learning algorithms, which are linear, deep learning algorithms are arranged in a hierarchy of increasing complexity and abstraction.

To comprehend deep learning, envision a toddler who utters the word "dog" as his first word. The toddler learns what a dog is -- and what it is not -- by pointing at various objects and uttering the word "dog." The parent responds by saying, "Yes, that is a dog," or, "No, that is not a dog." As the toddler continues to point at things, he becomes more aware of the characteristics that all dogs possess. The toddler is constructing a multi-level hierarchy, building upon knowledge acquired from each previous level, to create a sophisticated abstraction - the notion of a dog - all without being consciously aware of the process.

1.4: Practical Uses of Machine Learning in Real-life Scenarios

Using machine learning, computer systems can effectively process and leverage all available customer data. These systems follow predetermined protocols but can adjust to changing circumstances. By learning from the data, algorithms can produce new behaviours that were not initially planned.

Furthermore, with the ability to read and comprehend context, a digital assistant could effectively scan emails and extract key information. By utilizing this learning, the system can generate predictions about future client behaviour. This provides businesses with a deeper understanding of their customers, allowing them to be proactive rather than merely reactive.

Machine learning is a flexible technology with a wide range of uses in various disciplines and sectors. Here are six examples on how machine learning is getting utilized in real-life scenarios:

1. Image Recognition

Image recognition is a technology that allows machines to recognize and categorize objects within digital images or photographs. It is a form of computer vision, a field of study focused on enabling computers to interpret and understand digital visual information from the world around us. There are various applications where image recognition is utilized, such as security and surveillance, autonomous vehicles, medical diagnosis, and e-commerce.

Real-world applications of image recognition include:

- Face recognition: This is a common application of image recognition that is used for security and surveillance purposes, as well as for authentication on devices like smartphones.

- Autonomous vehicles: Autonomous vehicles, such as self-driving cars, employ image recognition technology to identify objects and individuals on the road. This aids the vehicle in determining when to accelerate, brake, or steer.

- Medical diagnosis: Image recognition is used in medical imaging to identify abnormalities or potential health risks in x-rays, MRI scans, and other medical images.

- E-commerce: E-commerce companies utilize image recognition technology to offer customers product recommendations based on their visual preferences.

- Agriculture: The agriculture industry employs image recognition technology to recognize plant diseases, pests, and other factors that impact crop health and productivity.

2. **Speech Recognition**
    - Another common application of machine learning in the real world is speech recognition.

It involves the translation of spoken words into text using software applications that can convert live or recorded speech into a text file. The intensity of time-frequency bands is used to segment the speech.

Real-world applications of speech recognition include:

- Smart Speakers: Smart speakers like Amazon Echo, Google Home, and Apple HomePod are all examples of speech recognition technology. These devices allow you to ask questions, play music, and control your smart home devices using just your voice.

- Voice Assistants: Many smartphones and tablets now come with built-in voice assistants that use speech recognition technology. Apple's Siri, Google Assistant, and Samsung's Bixby are all examples of voice assistants that can help you with tasks like setting reminders, sending texts, and making phone calls.

- Automated Phone Systems: Many businesses use speech recognition technology in their automated phone systems to help route calls and answer common questions. When you call a company's customer service hotline, you may be prompted to speak your request or problem, and the system will use speech recognition to route you to the appropriate department.

- Transcription Software: Speech recognition technology is often used in transcription software to transcribe voice messages or dictation into written text. This can be useful for people with disabilities or for anyone who wants to quickly dictate a message without having to type it out.

- Language Translation: Speech recognition technology can also be used to translate spoken language into another language in real-time. This technology can be particularly useful for people traveling to foreign countries or for international business communications.

3. **Medical Diagnose**

    - Medical diagnosis in machine learning refers to the utilization of AI algorithms to assist medical professionals in diagnosing medical conditions. Its algorithms can be trained to recognize patterns in large sets of medical data, such as patient symptoms, medical history, and test results, and use this information to make predictions about a patient's health.

    - In machine learning, medical diagnosis is usually done in two steps: training the algorithm and applying it to new data. To train the algorithm, medical professionals input large sets of medical data, including patient symptoms and outcomes, into the system. The data is then utilized by the algorithm to identify patterns and generate a model that can be used to make predictions about patients.

- Once the algorithm is trained, it can then assist medical professionals in diagnosing medical conditions. When a patient presents with symptoms, the medical professional inputs the patient's data into the system, and the algorithm uses its model to make a prediction about the patient's diagnosis. The medical professional can then use this prediction to inform their diagnosis and treatment plan.

Real-world applications of machine learning in medical diagnosis include:

- Cancer Diagnosis: Machine learning algorithms are used to analyze medical images, such as mammograms, X-rays, and CT scans, to detect signs of cancer. These algorithms can process these images much faster and more accurately than human radiologists, which can help catch cancers at earlier stages when they are more treatable.

- Diabetic Retinopathy: Diabetic retinopathy is a common complication of diabetes that can lead to blindness. ML algorithms can be used to analyze retinal images and detect signs of diabetic retinopathy. This can help doctors detect the condition earlier, leading to better outcomes for patients.

- Heart Disease: Machine learning algorithms can be trained to analyze patient data, including medical history, test results, and imaging data, to predict the risk of heart disease. This can help doctors identify high-risk patients and provide them with appropriate treatments and preventative measures.

- Alzheimer's Disease: Machine learning algorithms can analyze medical images, such as MRI scans, to detect signs of Alzheimer's disease. This can help doctors diagnose the condition earlier, allowing for earlier interventions and treatments to slow the progression of the disease.

- Skin Cancer: Machine learning algorithms can be trained to analyze images of skin lesions to catch signs of skin cancer. This can help dermatologists diagnose skin cancer earlier, leading to better outcomes for patients.

4. Statistical Arbitrage

- An additional instance of machine learning being applied to practical finance is statistical arbitrage, which utilizes an automated trading algorithm to examine a collection of securities based on economic variables in order to execute a trading strategy.

- By leveraging machine learning algorithms, traders can quickly and accurately process and analyze decent amounts of data, allowing them to make faster and more reliable trading decisions. Machine learning algorithms have the capability to adjust to varying market conditions, enabling traders to adjust their strategies in response to shifting market dynamics.

Real-world applications of statistical arbitrage include:

- High-Frequency Trading: Algorithms are frequently used to analyze large volumes of market data in real-time to find profitable trading possibilities, especially in high-frequency trading. These algorithms can quickly identify small differences in prices between similar financial instruments, allowing traders to make profitable trades within seconds.

- Portfolio Optimization: The optimization of investment portfolios can be achieved by utilizing machine learning algorithms to recognize overvalued or undervalued assets. This technique aids traders in identifying assets that are more likely to generate profits by analyzing market data, allowing them to adjust their portfolio accordingly.

- Risk Management: In this algorithm can be used to manage risks associated with trading by analyzing tremendous amounts of data and identifying potential risks. By using advanced statistical models and its algorithms, traders can identify potential risks in real-time and adjust their trading strategies to minimize losses.

- Fraud Detection: This technology detect fraud by analyzing good amounts of data and identifying patterns that may indicate fraudulent activity. By using advanced statistical models and machine learning algorithms, financial institutions can detect fraudulent activity early and take appropriate action to prevent losses.

5. **Predictive Analytics**
    - It consists of grouping available data into categories based on rules established by analysts

and computing the probability of an event or failure transpiring.

- Historical data is used to train machine learning algorithms in predictive analytics for identifying patterns, relationships, and make accurate predictions about future events. These algorithms use statistical models to recognize the most important attribute that affect the outcome of an event and then use this information to make predictions.

Real-world applications of predictive analytics include:

- Marketing: Predictive analytics is widely used in marketing to predict customer behavior and likings, optimize marketing campaigns, and improve customer engagement. By using this technology and analyze good amount of customer data, marketers can identify patterns and trends of the market that help them better understand their customers and deliver more targeted and personalized marketing campaigns.

- Finance: This tech is used in finance to identify credit risk, forecast stock prices, and optimize trading strategies. It can analyze vast amounts of financial data and identify patterns and trends that help traders and investors make informed decisions.

- Healthcare: This industry uses predictive analytics to identify patients who are showing signs of developing certain diseases, optimize treatment plans, and improve patient outcomes. It can also evaluate patient data and spot trends and patterns that help healthcare professionals make more accurate/reliable diagnoses and provide more personalized treatment.

- Manufacturing: Predictive analytics is used in manufacturing to optimize supply chain management, predict equipment failures, and improve production efficiency. On the other hand, it can analyse huge amounts of production data and identify patterns and trends that help manufacturers make informed decisions about production processes and equipment maintenance.

6. **Extraction**

- It is the process of identifying and capturing meaningful features or patterns from raw data. The goal of extraction is to transform the data into a format that can be easily understood and used by machine learning algorithms.

- Extraction is typically one of the initial steps in the machine learning process, and it can involve a variety of techniques, such as dimensionality reduction, feature selection, and feature engineering. These techniques are used to extract the most relevant information from the data and reduce the noise or irrelevant data that could adversely affect the precision of the machine learning model.

Real-world examples of extraction include:

- Text Mining: In natural language processing, extraction is used to identify and capture the most important information from unstructured data sources such as emails, social media posts, and web pages. This process involves extracting relevant features such as sentiment analysis, entity recognition, and topic modeling.

- Computer Vision: In image processing, extraction is used to identify and capture meaningful features from images, such as edges, shapes, textures, and colors. These features are then used to instruct machine learning algorithms for various tasks such as object detection, recognition, and classification.

- Signal Processing: In audio and speech processing, extraction is used to identify and capture relevant features from audio signals, such as frequency, amplitude, and duration. These features are then used build algorithms for speech recognition, audio classification, and other tasks.

- Finance: In finance, extraction is used to capture relevant information from financial data sources such as stock prices, financial statements, and news articles. This process gathers the information and the use it to predict future stock prices or market trends.

Overall, machine learning is a robust tool that can automate complex tasks and provide valuable insights in various fields and industries.

> ***Did you know?***
>
> AI has a language of its own: In 2017, researchers discovered that AI agents developed their own "language" to communicate with each other while being trained to negotiate. This fascinating development shows how AI can exhibit unexpected emergent behaviors.

1.5 What is Natural Language Processing (NLP)?

Natural language processing (NLP) is a branch of artificial intelligence that focuses on how computers and people can work together using language. It aspires to teach computers how to understand and produce human speech. NLP involves using a combination of computational linguistics, machine learning, and statistical models to analyze and derive meaning from human language data such as text, speech, and images.

In essence, NLP enables computers to understand human language and communicate with humans in a way that is natural and intuitive, similar to how humans communicate with each other. It finds extensive usage across multiple domains such as language translation, text summarization, sentiment analysis, chatbots, speech recognition, and more.. NLP is an exciting and rapidly evolving field that holds great promise for improving the way humans interact with machines, and vice versa.

Some of the key techniques used in NLP include:

- **Tokenization**

This is the process of splitting a text or a sentence into individual words or tokens. Tokenization is commonly used as an initial step in NLP before text is analyzed or processed further. The tokens created in this process can be analyzed in various ways, such as through part-of-speech tagging, sentiment analysis, or text classification.

Tokenization is important because computers do not understand natural language in the same way that humans do. To enable computers to process and analyze text, it needs to be changed into a format that they can understand, which is usually a series of numerical values. Tokenization helps achieve this by breaking down text into smaller units, each of which can be assigned a numerical value or code.

As an example, let's take a look at the following sentence: "The quick brown fox jumps over the lazy dog." In tokenization, this sentence would be broken down into individual words or tokens, like this:

1. "The"
2. "quick"
3. "brown"
4. "fox"
5. "jumps"
6. "over"
7. "the"
8. "lazy"
9. "dog"

Once the sentence has been tokenized, each token can be analyzed in various ways. For example, part-of-speech tagging can be utilized to identify the function of each word in the sentence (e.g., noun, verb, adjective). Text classification can be used to categorize the sentence based on its content, such as whether it is about animals or sports.

Tokenization is a crucial step in many NLP tasks and is used in a wide range of applications, including search engines, chatbots, sentiment analysis, and machine translation. Effective tokenization requires consideration of various factors, such as language-specific rules, context, and punctuation.

- **Part of Speech (POS) Tagging**

Part-of-Speech (POS) tagging is a natural language processing task that involves identifying and labelling the part of speech for each word in a given text or sentence. The goal of POS tagging is to assign each word a specific tag that indicates its grammatical role in the sentence, such as noun, verb, adjective, adverb, pronoun, preposition, conjunction, or interjection. This process is important in many language-related tasks, such as machine translation, information retrieval, and speech recognition.

Machine learning algorithms and statistical models trained on large corpora of annotated data are commonly used for POS tagging. In order to determine the most appropriate POS label for each word, these models take into account factors like context, morphology, and syntax. These models use various features, such as word context, morphology, and syntax, to predict the correct POS tag for each word. While POS tagging is a challenging task due to the ambiguity and variability of natural language, it has become an essential component of many modern language processing systems.

- For Example: "She gave me a book to read." In this sentence, "She" is a pronoun (PRP), "gave" is a verb (VBD), "me" is a pronoun (PRP), "a" is a determiner (DT), "book" is a noun (NN), "to" is a preposition (TO), and "read" is a verb (VB). The POS tagging for this sentence would be: PRP VBD PRP DT NN TO VB.

- **Named Entity Recognition (NER)**

Named Entity Recognition (NER) is a subtask of natural language processing that involves identifying and categorizing named entities in text. Named entities are words or phrases which are associated with specific individuals, locations, organizations, products, or other entities that are referred to by a proper name. The ultimate goal of NER is to detect and classify these entities automatically in a given text and assign them to pre-defined categories such as person, location, organization, date, time, product, or event. This process is important for many language-related applications, such as information extraction, machine translation, sentiment analysis, and question answering.

As an illustration, let's take a look at the following sentence: "John works for Google in New York." In this sentence, "John" is a person, "Google" is an organization, and "New York" is a location. A NER system would identify these entities and assign them to the appropriate categories where the belong, resulting in the following output:

John (person) works for Google (organization) in New York (location).

NER can be performed using a variety of techniques, such as rule-based systems, statistical models, or deep learning algorithms. These systems rely on various features, such as context, syntax, and morphology, to figure out and classify named entities in text. While NER is a challenging task due to the complexity and variability of natural language, it has become an essential component of many modern language processing systems.

- **Stemming and Lemmatization**

These are two common techniques used in NLP to reduce words to their base or root form. The goal of both techniques is to simplify the vocabulary and increase the efficiency and accuracy of language processing algorithms. While both techniques aim to reduce words to their base forms, they do so in different ways.

Stemming involves reducing words to their stem or root form by removing suffixes and prefixes. For example, the stem of the word "running" would be "run," and the stem of the word "dogs" would be "dog." Stemming algorithms use various rules and heuristics to identify the stem of a word, and different stemming algorithms may produce different results. While stemming is a relatively easy and fast technique, it can sometimes result in incorrect stem words due to the overzealous removal of suffixes and prefixes.

Lemmatization, on the other hand, involves reducing words to their base form by analyzing their morphological structure and context. For example, the lemma of the word "running" would be "run," and the lemma of the word "dogs" would be "dog." Lemmatization algorithms use various linguistic rules and dictionaries to identify the lemma of a word, and they can handle irregular words and inflections better than stemming algorithms. However, lemmatization is generally very slow and complex than stemming.

Let's take the following sentence as an example: "The dogs are running in the park." A stemming algorithm would produce the following output: "The dog are run in the park," while a lemmatization algorithm would produce the following output: "The dog be run in the park." In this example, the stemming algorithm produced some incorrect results, while the lemmatization algorithm produced more accurate results.

- **Sentiment Analysis**

Sentiment Analysis is a natural language processing technique that involves automatically identifying and categorizing the emotional tone or attitude expressed in a piece of text. The goal behind this is to classify the polarity of the text as positive, negative, or neutral. Sentiment Analysis is useful in a variety of applications, such as social media monitoring, customer feedback analysis, and brand reputation management.

To demonstrate, let's examine the following sentence: "I absolutely loved the new movie - it was amazing!" In this sentence, the sentiment is clearly positive. A sentiment analysis system would identify the positive sentiment and classify the sentence as positive. On the contrary, let's consider the following sentence: "I was very disappointed with the service at the restaurant." In this sentence, the sentiment is negative, and a sentiment analysis system would identify the negative sentiment and classify the sentence as negative.

Sentiment Analysis are able to performed using a variety of techniques, such as lexicon-based approaches, machine learning algorithms, or deep learning models. These systems rely on various features, such as word frequency, context, and syntax, to identify the sentiment expressed in a piece of text.

- **Topic Modeling**

Its is very statistical technique used to uncover the hidden semantic structures within a large corpus of text data. It is a method of discovering abstract topics or themes that run through a collection of documents or text data sets. In essence, it is a form of exploratory analysis that seeks to find the patterns and themes that emerge from a set of documents, based on the frequency of specific words or phrases. The technique uses algorithms such as Latent Dirichlet Allocation (LDA) or Non-negative Matrix Factorization (NMF) to cluster words and phrases that frequently co-occur in the text. These clusters are then used to identify topics that run through the documents.

Topic modeling has been widely used in many fields such as social sciences, natural language processing, and digital humanities. It has proved to be particularly useful in analyzing large corpora of unstructured text data, such as social media posts, news articles, and scientific papers. It can provide insights into the main themes or trends that emerge from such data, enabling researchers to identify patterns and make informed decisions. Topic modeling is also useful in machine learning and text mining applications, such as sentiment analysis, recommendation systems, and text classification.

Here are some examples of how topic modeling has been applied in different fields:

- Social media analysis: Topic modeling has been used to analyze social media posts and tweets to understand public opinion on different topics, such as politics, sports, and entertainment. For example, researchers have employed topic modeling to analyze tweets related to the 2016 US presidential election to identify key issues and sentiments expressed by Twitter users.

- Healthcare: The analysis of electronic medical records using topic modeling has been applied to identify patterns and trends in patient data. For example, researchers have used this technology to identify topics related to patient satisfaction, medication adherence, and disease management.

- Digital humanities: The use of topic modeling to analyze large collections of literature and historical documents has been implemented to uncover trends and themes. An illustration of the use of topic modeling is the analysis of Shakespeare's works to identify repetitive themes and motifs.

- News analysis: Topic modeling has been utilized in the examination of news articles to recognize noteworthy topics and trends present in the media. For example, researchers have used topic modeling to analyze news articles related to climate change to identify the main themes and opinions expressed by journalists and experts.

- Overall, this is a strong tool for understanding and organizing large sets of unstructured text data, providing a framework for analyzing and extracting insights from complex data sets.

- **Dependency Parsing**

Dependency Parsing is a subdomain of NLP that concentrates on analyzing the grammatical organization of a sentence by identifying the dependencies or relationships among words. In this approach, the words in a sentence are represented as nodes in a graph, and the relationships between them are represented as directed edges that show the dependency between them. Dependency Parsing is used to analyze the structure of sentences in many

NLP applications such as machine translation, information extraction, and sentiment analysis.

In Dependency Parsing, the agenda is to pinpoint the head or governor word in a sentence, which is the main word that governs the syntactic structure of the sentence. Other words in the sentence are dependent on the head word, and the relationships between them are represented as arcs. These arcs indicate the type of dependency between the words, such as subject-verb, object-verb, modifier-noun, and so on. The analysis of the sentence structure is typically represented as a tree, where the head word is the root of the tree, and the arcs represent the edges.

There are two main approaches to Dependency Parsing: Transition-based Parsing and Graph-based Parsing. In Transition-based Parsing, a parsing algorithm moves through the sentence from left to right, and incrementally builds the tree structure by making a series of decisions at each step. In Graph-based Parsing, the parser creates a graph of the sentence, and then searches for the best tree structure that satisfies certain constraints.

Dependency Parsing has a broad spectrum of applications in NLP, including machine translation, named entity recognition, information extraction, and sentiment analysis. It is an important tool for understanding the structure of language, and for developing more sophisticated NLP models that can accurately capture the nuances of natural language.

> ### *Did you know?*
> AI can generate its own jokes: AI systems have been trained to understand humor and even create jokes, albeit with varying degrees of success. This demonstrates AI's potential to grasp complex aspects of human communication and expression.

CHAPTER 2
COMPUTER VISION

Computer vision refers to a field of artificial intelligence that allows machines to derive significant information from digital images, videos, and other visual data. This technology helps computers and systems to "see" and "understand" visual inputs, ultimately making recommendations and executing actions depending on the information gathered. Although computer vision operates similarly to human vision, there are distinct differences. While humans have a lifetime of experiences to learn and understand how to differentiate between objects, perceive distance, movement, and anomalies, computer vision must use cameras, algorithms, and data to replicate these functions in a much shorter amount of time. For instance, a system trained to monitor production assets or inspect products can evaluate thousands of objects per minute, swiftly detecting imperfections or issues that may be unnoticeable to human observers. Thus, computer vision has the potential to surpass human capabilities in certain visual tasks.

2.1 Image Recognition

Image recognition is a broad term that refers to the ability of machines to interpret and understand the content of an image. It encompasses a range of computer vision techniques, including object recognition, image classification, and image segmentation.

Object recognition involves identifying and labelling objects in an image, while image classification requires categorizing an image into one of several predefined classes or categories. To accomplish this, the image is partitioned into numerous layers or regions, with each one corresponding to a specific element or section of the image

Image recognition is a fundamental task in many computer vision applications, such as autonomous driving, medical imaging, and security and surveillance. It is a challenging task due to the variability of image content and the need to account for occlusions, background clutter, and object variability.

In order to attain elevated levels of precision in recognizing images, scientists frequently employ advanced deep learning methods, like convolutional neural networks (CNNs). These networks are capable of automatically identifying intricate features and patterns from images and videos, and have demonstrated exceptional success in image recognition challenges by outperforming other techniques and achieving state-of-the-art results on various benchmark datasets.

THE POWER OF AI

Despite the progress made in image recognition, there are still many challenges to be addressed, such as improving the robustness of systems to changes in lighting and perspective, and developing more efficient algorithms that can run in real-time on low-power devices.

Here's an example of image recognition in the context of autonomous driving:

One of the key tasks in autonomous driving is to detect and classify objects in the environment, such as pedestrians, other vehicles, and traffic signs. To achieve this, a camera or a set of cameras mounted on the vehicle can capture images of the surrounding environment, and an image recognition system can analyze these images in real-time to identify and track objects of interest.

The image recognition system typically consists of a convolutional neural network that is trained on a large dataset of labeled images, such as the COCO (Common Objects in Context) dataset. During training, the CNN learns to recognize the features and patterns in the images that correspond to each object class, such as the shape and texture of a pedestrian, or the color and shape of a traffic sign.

Once trained, the CNN can be deployed on the autonomous vehicle to analyze images captured by the cameras in real-time. The CNN processes each image and identifies the location and type of objects in the scene, such as the location and size of a pedestrian or the type of traffic sign present.

The output of the image recognition system can then be utilized to inform the decision-making process of the autonomous driving system. For example, if a pedestrian is detected in the path of the vehicle, the system can apply the brakes to avoid a collision.

Overall, image recognition is a critical component of autonomous driving and is essential for ensuring the safety and reliability of these systems in real-world scenarios.

2.2 Object Detection

Object detection is a computer vision technique that involves detecting and localizing objects in an image or video. It goes beyond image classification, which only categorizes an image into one of several different predefined classes or categories, by identifying the specific location of each object in the image.

Object detection can be achieved using a variety of techniques, such as template matching, sliding window detection, and region proposal-based methods. One common approach to object detection is to use a type of neural network called a region-based convolutional neural network (R-CNN).

R-CNNs are designed to first generate a set of region proposals, which are potential object locations in the image. The network then extracts features from each proposed region and uses these features to classify the region and refine its position if necessary.

There are two types of object detection:

- Two-stage object detection: This involves generating a set of region proposals using a separate algorithm or network, and then using a CNN to extract features from each proposed region and perform object classification and localization.

- One-stage object detection: This involves directly predicting object classes and locations in a single pass through the network, without the need for separate region proposal generation.

Object detection is a fundamental task in many computer vision applications, such as surveillance, robotics, and autonomous driving. It is a challenging task due to the variability of object appearance, shape, and size, as well as the need to handle occlusions and background clutter.

State-of-the-art object detection systems can achieve high accuracy rates on some datasets, but there is still much research to be done to improve the performance of these systems in real-world scenarios. For example, there is ongoing research to improve the speed and efficiency of object detection algorithms to enable real-time processing on low-power devices.

For Example, in a surveillance system, the agenda is to detect and track objects of interest, such as people or vehicles, in a scene captured by a camera or set of cameras. Object detection can be used to detect these objects and track their movements over time automatically.

To achieve this, a computer vision algorithm can be trained on a large dataset of labeled images, such as the COCO (Common Objects in Context) dataset, to learn to recognize the features and patterns in images that correspond to each object class. The algorithm can then be used to analyze frames from the surveillance video and detect and track objects in real-time.

One popular approach to object detection in surveillance is the YOLO (You Only Look Once) algorithm. Designed to be fast and efficient, it is a one-stage object detection algorithm that is highly suitable for real-time applications. It works by dividing the image into a grid of cells and predicting the probability of an object being present in each cell, as well as the class of the object and its bounding box.

Once objects have been captured and localized in each frame of the video, they can be tracked over time using techniques such as Kalman filters or particle filters. This allows the surveillance system to monitor the movements of objects of interest and identify any suspicious or anomalous behavior.

2.3 Image Segmentation

Image segmentation is a computer vision technique that involves splitting an image into multiple segments, or regions, based on the content of the image. Each segment typically corresponds to a distinct object or region of interest in the image.

Segmentation can be accomplished at different levels of granularity, spanning from basic foreground-background differentiation to more intricate tasks such as semantic segmentation, instance segmentation, and panoptic segmentation.

Here are some examples of image segmentation tasks:

- Semantic segmentation: This involves dividing an image into multiple segments and assigning each segment a label corresponding to the semantic category of the objects or regions it contains. For example, a semantic segmentation algorithm might label one segment as "person", another segment as "car", and another segment as "building".

- Instance segmentation: This involves not only assigning each segment a semantic label, but also distinguishing between different instances of the same object. For example, an instance segmentation algorithm might label each individual in an image with a unique identifier, allowing it to track each person's movements and interactions.

- Panoptic segmentation: This involves combining semantic and instance segmentation to create a unified segmentation map that covers all objects and regions in the image. This allows for a more complete and detailed understanding of the content of the image.

Image segmentation is a critical component of many computer vision applications, such as object recognition, autonomous driving, and medical imaging. It is a very difficult task due to the variability of object appearance, shape, and size, as well as the need to handle occlusions and background clutter.

State-of-the-art segmentation algorithms typically rely on deep learning techniques, such as

CNN, and are highly trained on large datasets of labeled images. They can achieve high accuracy rates on some datasets, but there is still much research to be done to enhance the performance of these algorithms in real-world scenarios.

Image segmentation is used in a wide range of real-life applications, including:

- Medical Imaging: Medical professionals use image segmentation to identify and separate different structures in medical images such as CT scans, MRI scans, and X-rays. This helps in accurate diagnosis and treatment of various diseases and conditions.

- Autonomous Vehicles: Image segmentation allows the vehicle to make decisions based on its environment and avoid collisions.

- Robotics: Image segmentation is used in robotics to identify and track objects in a robot's environment. This can be used to improve the accuracy of robot navigation and manipulation tasks.

- Surveillance and Security: Image segmentation is used in surveillance systems to detect and track objects of interest, such as people and vehicles, in real-time. This allows security personnel to quickly identify and respond to potential threats.

- Augmented Reality: Image segmentation is used in augmented reality applications to separate the foreground and background in an image or video. This allows virtual objects to be placed on top of real-world objects and provides a more immersive experience for the user.

- Video Editing: Image segmentation is used in video editing software to separate the foreground and background of a video clip. This allows editors to apply effects and make changes to the foreground and background separately, improving the quality of the final video.

Overall, image segmentation is a powerful technology that has numerous applications in many different industries and fields.

2.4 Video Analysis

Video analysis is a computer vision method that encompasses the examination and comprehension of a video sequence's content. It can include tasks such as object detection, tracking, recognition, segmentation, and activity recognition.

Here are some examples of video analysis tasks:

- Object detection and tracking: This involves identifying and tracking objects of interest in a video sequence over time. For example, a video analysis algorithm might track the movement of a person or a vehicle as it moves through a scene.

- Action recognition: This involves identifying and classifying the actions being performed by multiple objects in a video sequence. For example, a video analysis algorithm might be able to recognize when a person is walking, running, or jumping.

- Event detection: This involves detecting and classifying events that occur in a video sequence, such as accidents, fights, or fires. This can be utilized for surveillance or security applications.

- Video segmentation: This involves dividing a video sequence into multiple segments based on the content of the video. For example, a video analysis algorithm might segment a sports game into different parts based on the different plays and events that occur.

- Emotion recognition: This involves recognizing the emotional state of individuals in a video sequence, such as happiness, sadness, or anger. This can be useful in fields such as psychology or marketing.

Video analysis is used in many real-life applications, such as:

- Security and surveillance: Video analysis is used in security and surveillance systems to detect and track individuals and objects of interest in real-time. This can help to improve public safety, prevent crime, and provide evidence for investigations.

- Entertainment: Video analysis is used in the entertainment industry to analyze and improve performances in sports and other competitions. For example, coaches and athletes use video analysis to study their movements and technique in order to make improvements.

- Education and training: Video analysis is used in education and training to provide feedback and improve learning outcomes. For example, teachers might use video analysis to evaluate a student's presentation skills or to analyze a classroom discussion.

- Medical and healthcare: Video analysis is used in medical and healthcare applications to analyze human movement, such as in physical therapy and rehabilitation. It is also very useful to monitor patients remotely or to detect signs of medical conditions such as Parkinson's disease.

- Traffic and transportation: Video analysis is used in traffic and transportation systems to monitor traffic flow, detect accidents and incidents, and improve safety on roads and highways.

Overall, video analysis is an important technology that is used in many different fields and industries. It can provide valuable insights and understanding of complex events and situations, and can help to automate tasks that would otherwise be difficult or time-consuming.

CHAPTER 3
AI IN ACTION

3.1: AI in Healthcare

AI is revolutionizing healthcare in a variety of ways. The main benefit of AI is its ability to process vast amounts of data quickly and accurately. This is particularly useful in healthcare, where there is a wealth of patient data that can be analyzed to improve diagnosis, treatment, and patient outcomes.

One area where AI is making a big impact is in medical imaging. AI algorithms can analyze medical images like X-rays, CT scans, and MRIs to detect anomalies and help doctors make more accurate diagnoses. For example, AI can identify signs of cancer in a mammogram or detect fractures in an X-ray with high accuracy, potentially improving patient outcomes and reducing the need for invasive procedures.

AI is actively involved in the development of personalized treatment plans based on a patient's unique medical history and genetic makeup. By analyzing patient data, AI algorithms can identify patterns and make predictions about which treatments are most likely to be effective for a particular patient. This can help doctors develop targeted treatment plans that are more effective and have fewer side effects.

Another area where AI is making a big impact is in drug discovery. By employing AI algorithms, it becomes feasible to sift through enormous datasets to identify potential drugs that may be effective and can even simulate how drugs will interact with the body. This can help researchers develop new drugs more quickly and efficiently, potentially leading to faster development of treatments for diseases like cancer and Alzheimer's.

AI is also being used to improve patient care and outcomes through predictive analytics. By analyzing patient data, AI algorithms can pinpoint patients who are at high risk for complications or readmissions, allowing doctors to intervene early and prevent adverse outcomes.

Without a doubt, artificial intelligence is transforming the medical field by improving patient diagnoses, therapies, and overall outcomes. While there remain various challenges to tackle, including ensuring the ethical and responsible application of AI, it holds the capacity to bring about a paradigm shift in healthcare results.

Did you know?

AI supports mental health: AI-driven chatbots and virtual therapists can provide mental health support by engaging users in conversation, tracking mood patterns, and offering personalized coping strategies, making mental health resources more accessible.

3.2: AI in Finance

The finance industry is experiencing a significant impact from AI. One of its most significant advantages in finance is its capacity to swiftly and accurately analyze large volumes of data, rendering it an effective instrument for detecting fraud, anticipating market trends, and enhancing decision-making.

AI is significantly impacting fraud detection as one of its most significant applications. Its algorithms have the ability to examine large amounts of financial data to detect unusual patterns or behavior that may indicate fraud. This can help financial institutions identify and prevent fraudulent activity before it causes significant financial losses.

The application of AI is also enhancing risk management in finance. Its algorithms can analyze significant amounts of data, detect patterns, and assist financial organizations in evaluating risk and making informed investment decisions. This can potentially lead to higher returns and promote a more secure financial system overall.

Another area where AI is making a big impact is in customer service. With the help of AI, chatbots can manage fundamental customer inquiries and offer personalized recommendations based on the individual's financial history and preferences. This can improve customer satisfaction and reduce the workload for human customer service agents.

Managing investment is simple as ABC for AI and its now being employed to advance investment management. How? By analyzing large amounts of financial data, AI algorithms can identify trends and make predictions about market performance. This can help investors make more informed decisions and achieve better returns on their investments.

3.3 AI in Transportation

The transportation industry is undergoing a revolution in various aspects due to the incorporation of AI, which ranges from augmenting safety and efficacy to enhancing the overall travel experience of commuters. One area where AI is making a significant impact is in the expansion of self driving vehicles. By incorporating AI algorithms and sensors, these vehicles are able to navigate roads and traffic, avoid collisions, and make appropriate decision in real-time.

In addition to self-driving cars, AI is also being used to optimize traffic flow and reduce congestion on roadways. By collecting and analyzing data from traffic cameras, GPS devices, and other sources, AI algorithms are able to identify traffic patterns and make real-time adjustments to traffic lights and other

THE POWER OF AI

infrastructure to reduce delays and improve the flow of vehicles.

AI is being leveraged to improve safety in transportation. For example, many vehicles are equipped with advanced safety features such as collision avoidance systems, lane departure warnings, and adaptive cruise control that use AI to monitor the vehicle's surroundings and respond appropriately to potential hazards.

In the aviation industry, it also improves the safety and efficiency in air traffic control. By analyzing data from radar and other sensors, AI algorithms can help air traffic controllers make well considered decision about routing and scheduling aircraft, which can help reduce delays and improve safety.

Finally, AI is also being used to improve the overall travel experience for passengers. For example, airlines and other transportation providers are using AI chatbots to provide customers with real-time information about their travel plans, answer questions, and resolve issues quickly and efficiently. we can expect to see even more innovations and improvements in transportation in the years to come.

3.4: AI in Retail

Artificial intelligence is metamorphosing the retail industry in many ways, from optimizing inventory management to improving customer experiences. One keyway AI is helping retail businesses is through the applications like chatbots and virtual assistants. These AI-powered tools are being used to help customers find products, answer questions, and complete transactions, all in a personalized and efficient manner.

Another area where AI is making a significant impact is in supply chain management. AI algorithms are being used to analyze data from sales, customer demand, and inventory levels to optimize ordering, forecasting, and restocking. This allows retailers to reduce waste, save money, and ensure they always have the right products in stock.

AI is also being used to enhance the in-store experience for customers. For example, some retailers are using AI-powered cameras to track customer behavior and analyze store layouts to optimize product placement and improve store design. This can help retailers create more inviting and user-friendly shopping environments that encourage customers to spend more time in the store and make more purchases.

Additionally, AI is helping retailers personalize their marketing efforts. By studying customer data, AI algorithms can identify patterns and preferences and create personalized recommendations and targeted advertising. This helps retailers better

THE POWER OF AI

connect with customers and improve their overall shopping experience.

Finally, to improve fraud detection and security in the retail industry AI is playing very important role. Also, by gathering customer data and transactions, AI algorithms can identify suspicious behavior and flag potential fraud, which can help retailers protect themselves and their customers from financial losses.

3.5: AI in Manufacturing

AI (Artificial Intelligence) in manufacturing refers to the application of machine learning algorithms and other advanced analytical techniques to automate and optimize the various processes involved in manufacturing. This includes everything from assembly line operations, to supply chain management, to quality control.

AI in manufacturing enables companies to leverage data from sensors, machine logs, and other sources to gain insights into their operations, identify patterns, and optimize processes in real-time. This can drastically help business to reduce costs, improve quality, increase production efficiency, and enhance overall productivity.

Some examples of AI applications in manufacturing include:

- Predictive maintenance

To predict when equipment will require maintenance, AI application can predict that for employer and provide detailed information. This allows manufacturers to schedule maintenance proactively, before a failure occurs, reducing downtime and repair costs. Predictive maintenance systems typically rely on data from sensors and other sources to monitor equipment health, detect anomalies, and identify patterns that indicate impending failures. By accumulating and analyzing this data in real-time, machine learning algorithms can generate predictive models that can help manufacturers optimize their maintenance schedules and minimize disruptions to production. This approach is particularly useful for expensive, critical, and hard-to-replace equipment, as well as for manufacturing environments that require high levels of reliability and uptime. Overall, predictive maintenance is an important application of AI in manufacturing that can help manufacturers to improve their productivity, reduce costs, and enhance their overall operational efficiency.

- Quality control

This is another important application of AI in manufacturing that involves using computer vision and machine learning algorithms to detect defects in products and ensure that they meet quality standards. Quality control systems can analyze images and video data from cameras and sensors to identify defects such as scratches, dents, or variations in color and shape. Machine learning algorithms can learn from this data to improve their accuracy and efficiency over time. In intricate and high-volume manufacturing processes, AI-powered quality control systems are highly beneficial for identifying defects. This is especially useful since manual inspection can be prone to errors and take up significant time. By automating the inspection process, AI can help manufacturers minimize the risk of product recalls and enhance their customer satisfaction levels. Moreover, detecting defects early in the manufacturing process can assist manufacturers in minimizing waste and enhancing production efficiency. In summary, AI-based quality control is a crucial manufacturing application that helps manufacturers guarantee their products adhere to the highest quality standards while simultaneously reducing expenses and improving productivity.

- Supply chain optimization

Supply chain optimization is a critical application of AI in manufacturing that involves using predictive analytics to optimize inventory levels, reduce lead times, and improve supply chain efficiency. By analyzing historical data on demand, production, and inventory, machine learning algorithms can generate predictive models that can help manufacturers optimize their inventory levels and improve their production planning. These models can help manufacturers to avoid stockouts and overstocks, reduce the cost of holding inventory, and improve their order fulfillment rates. Additionally, AI can be used to optimize logistics and transportation by predicting delivery times, optimizing routing, and reducing transportation costs. By improving supply chain efficiency, AI can aid manufactures reduce cost, improve their response times to changes in demand, and enhance their overall competitiveness.

- Autonomous robots

Autonomous robots are another important application of AI in manufacturing that involves using robots equipped with AI algorithms to fully automate tasks such as material handling, assembly, and inspection. These robots are typically equipped with sensors and cameras that allow them to perceive their environment, and machine learning algorithms that enable them to make decisions based on this data. By automating these tasks, autonomous robots can help manufacturers reduce costs, increase efficiency, and improve the quality of their products. For example, in material handling, robots can be utilized to transport raw materials and finished products between different locations, lowering down the need for human labor and increasing throughput. In assembly, robots can be used to assemble components with high precision and speed, reducing the likelihood of errors and refining the overall quality of the finished product. In inspection, robots can be used to perform quality checks on products, identifying defects that may not be noticeable to the human eye. Overall, autonomous robots are an important application of AI in manufacturing that can help manufacturers to improve their productivity, reduce costs, and enhance the quality of their products.

3.6 AI in Marketing

A growing number of businesses, along with their marketing teams, are quickly embracing intelligent technology solutions in order to enhance their operational efficiency and elevate the overall customer experience. One ultimate type of such solutions is AI marketing platforms. These platforms enable marketers to gain a more refined and thorough understanding of their target audiences. By leveraging insights generated through AI algorithms, marketers can drive conversions more effectively while simultaneously reducing the workload for their teams.

Here are some examples of how AI is changing the marketing business:

- Personalization

AI algorithms are being used to scrutinize customer data to personalize marketing campaigns. For example, Netflix uses AI algorithms to recommend movies and TV shows to its clients based on their preferences and viewing history in the past.

- Predictive Analytics

AI algorithms are being used to analyze large data sets to predict customer behavior. For example, Amazon uses AI algorithms to predict which products a customer is likely to purchase based on their search and purchase history.

- Chatbots

AI-powered chatbots are being used to provide 24/7 customer support and streamline the purchase process. For example, H&M uses a chatbot to help customers find products and make purchases through Facebook Messenger.

- Image and Video Recognition

AI algorithms are being used to study images and videos to identify specific products, people, or locations. For example, Pinterest uses AI algorithms to identify products in photos and provide users with links to purchase those products.

- Sentiment Analysis

AI algorithms are being used to monitor social media and online content to gauge customer sentiment and pinpoint potential issues. For example, Coca-Cola uses sentiment analysis to monitor social media and respond to customer complaints and feedback.

- Content Creation

AI algorithms are being used to create personalized content for customers. For example, The Washington Post uses AI algorithms to create personalized news articles for individual users based on their interests.

In short, AI is transforming the marketing business by providing businesses with new tools to connect with their customers, streamline processes, and make data-driven decisions.

CHAPTER 4
THE ROLE OF AI IN ROBOTS

THE POWER OF AI

Artificial intelligence plays a crucial role in robots. It is what allows robots to perform tasks that would otherwise require human intelligence and decision-making skills.

AI plays a crucial role in enabling robots to sense and comprehend their surroundings, which allows them to make informed choices and take actions accordingly. This involves using sensors such as cameras, microphones, and other detectors to gather data about the robot's surroundings, and then processing that data to identify objects, people, and other relevant information.

Once the robot has perceived its environment, AI algorithms can also be used to make decisions about what actions the robot should take. For example, a robot in a manufacturing plant might use AI to identify the optimal path for moving materials around the plant, while a robot in a hospital might use AI to navigate through hallways and find specific rooms.

In addition, AI is frequently utilized to enable robots to learn and adapt to new situations. This involves using ML to analyze data from the robot's sensors and adjust the robot's behavior accordingly. For example, a robot that is tasked with sorting objects might use machine learning to boost its accuracy over time by analyzing the results of its previous sorting attempts.

> **Did you know?**
>
> The term "robot" comes from the Czech word "robota," which means "forced labor" or "work." The term was first used in a 1920 play by Karel Čapek called "R.U.R. (Rossum's Universal Robots)."

4.1 What is Robotics?

Robotics is the field of technology that involves the design, construction, and operation of robots. A robot is a machine or device that can perform a variety of tasks automatically, with some level of autonomy, often with the help of programming and artificial intelligence.

Robots can be designed to perform a wide range of tasks, from simple repetitive tasks like assembling products on an assembly line, to more complex tasks like performing surgery or exploring other planets. Robotics combines multiple disciplines, including mechanical engineering, electrical engineering, computer science, and artificial intelligence, to design, build, and operate robots.

Robotics is a growing field with many applications in areas such as manufacturing, healthcare, defense, and space exploration. As technology advances, robots are becoming increasingly sophisticated and are capable of performing more complex tasks with greater accuracy and efficiency.

Overall, robotics is an interdisciplinary field that involves the use of technology to design and build machines that can perform a variety of tasks autonomously, improving efficiency and safety in many industries.

4.2 Are Robotics and AI the Same Thing?

Robotics and AI are not the same thing, although they are closely related and often used together in the same systems.

Robotics is a field of engineering that focuses on the design, construction, and operation of robots. Robots are physical machines that can be programmed to perform tasks autonomously or semi-autonomously, often with the help of sensors and actuators. Robotics involves the use of mechanical, electrical, and software engineering to create robots that can interact with the physical world and perform specific tasks.

On the other hand, the field of AI within computer science, is centered around developing machines with the ability to perform tasks that traditionally necessitate human intelligence, such as perception, reasoning, and decision making. AI involves the use of machine learning, deep learning, and other strategies to create algorithms that can analyze data, learn from it, and make predictions or decisions based on that data.

While robotics and AI are different fields, they are frequently used together to create intelligent robots that can perform complex tasks. For example, an autonomous car may use robotics to control its physical movements, while AI algorithms analyze data from its sensors to make decisions about how to navigate the road. Similarly, a manufacturing robot may use robotics to control its movements and AI to analyze data from sensors to optimize its performance.

4.3 What is The Role of AI in Robotics?

AI plays an essential role in robots by providing the ability to perceive, reason, and act in complex and dynamic environments. Here are few examples of how AI is being used in robots:

- Perception: AI algorithms are used to enable robots to perceive their environment using various sensors such as cameras, LIDARs, and sonars. AI can be used to detect objects, recognize faces, identify obstacles, and even understand natural language. For example, autonomous cars use AI to analyze visual data from their cameras and LIDARs to detect objects on the road and navigate safely.

- Planning and decision making: AI is used to enable robots to plan and make decisions based on their perception of the environment. AI algorithms can also be utilized to develop strategies, optimize paths, and make predictions based on sensor data. For example, a robot in a warehouse can use AI to optimize its path to pick up items in the most efficient way possible.

- Learning and adaptation: AI can help robots to learn from their experiences and adapt to new situations and machine learning can analyze data from sensors and adjust the robot's behavior accordingly. For example, a robot that is tasked with sorting objects can use machine learning to enrich its accuracy over time by analyzing the results of its previous sorting attempts.

- Interaction with humans: AI algorithms have the capability to assist robots in interacting with humans in a natural manner by comprehending natural language and identifying emotions. Social robots, for instance, have the ability to detect emotions, recognize faces, and react to human speech through AI, making them beneficial in healthcare and education.

In short, AI is a critical component of robots, enabling them to perceive, reason, and act in complex and dynamic environments.

4.4 Some Examples of AI applied to Robotics.

- Boston Dynamics' Spot Robot: This robot is equipped with AI algorithms that enable it to navigate through rough terrain and perform various tasks such as inspection, surveillance, and delivery. Its AI algorithms allow it to perceive its environment using cameras and sensors and make decisions about how to navigate and interact with its surroundings.

- Amazon's Kiva Robots: These robots are used in Amazon's warehouses to move shelves and products around the facility. The robots use AI algorithms to optimize their paths and avoid obstacles, making the warehouse more efficient and reducing the workload on human workers.

- iRobot's Roomba Vacuum: This robotic vacuum cleaner uses AI algorithms to map its environment and navigate around obstacles.

The Roomba also uses machine learning to adapt to different floor surfaces and cleaning preferences, making it more efficient over time.

- SoftBank Robotics' Pepper Robot: This social robot uses AI to interact with humans in natural ways, such as recognizing faces, understanding natural language, and detecting emotions. Pepper is used in various applications, including customer service, education, and healthcare.

- ABB Robotics' YuMi Robot: This robot is designed to work alongside humans in manufacturing environments. Its AI algorithms enable it to perceive and respond to its human co-workers, making it easier to collaborate on tasks and increasing overall productivity.

As AI technology advances, we can anticipate the emergence of more advanced robots that utilize AI to accomplish intricate tasks, in addition to the aforementioned applications.

CHAPTER 5
AI AND SOCIETY

5.1 The Impact of AI on Employment

The way we work, and its impact on the future of work is expected to be significant. AI technologies, such as ML and NLP, are presently employed to automate repetitive tasks, improve decision-making, and generate novel employment opportunities. While some jobs may be replaced by machines, there will be a growing demand for workers who can collaborate with AI systems and manage complex tasks.

The fusion of AI into the workplace will require workers to learn new skills and adapt to new roles. They will have to be proficient in data analysis, machine learning, and other AI technologies. This will create a need for retraining and upskilling programs that can help workers develop these skills. The rise of AI will also require employers to develop new policies and strategies to ensure that their workforce remains productive and engaged.

The implementation of AI in the workplace also raises important ethical and social issues which we will discuss later. It is essential to consider the impact of AI on job displacement, privacy, and fairness. Governments, businesses, and individuals need to work together to ensure that AI is used responsibly and ethically. The future of work will be shaped by AI, and it is up to us to ensure that this transformation is beneficial for all.

5.2 The Fusion of AI in Education

The potential impact of AI on the future of education is immense. Educators can leverage personalized learning experiences for students and improve learning outcomes. Benefits of AI in education is that it allows for the customization of teaching methods according to individual student needs and learning styles.

It can also help educators to better understand student progress and challenges, by providing insights into learning patterns, and identifying areas where students may need extra support. This can help educators to create more targeted and effective interventions to support student learning.

Another way AI is transforming education is through the creation of intelligent tutoring systems (ITS). ITS use ML to understand the way students learn and provide personalized feedback and guidance to help them improve their skills. These systems can be particularly useful in helping students who may be struggling with certain topics or concepts, as they can offer additional support in a way that is tailored to the student's individual needs.

AI can also be used to create immersive and engaging learning experiences, such as virtual and augmented reality applications. These technologies can help students to better understand complex concepts by providing interactive and visual learning experiences.

However, the integration of AI in education also raises concerns around privacy, security, and ethics. It is important for educators to be aware of these concerns and work to ensure that AI is used responsibly and in the best interests of students.

Overall, AI has the potential to revolutionize the way we teach and learn, but it is important to approach this technology with caution and to ensure that it is used in a responsible and ethical manner.

5.3 Revolution of Healthcare with AI

The potential of AI to transform the healthcare industry is enormous. By leveraging AI, healthcare practitioners can enhance patient results, simplify procedures, and anticipate and avert diseases in advance.

The capacity of AI to enhance diagnosis and treatment makes it one of the most substantial advantages of implementing this technology in healthcare. Its algorithms have the ability to examine vast quantities of patient data, including medical histories, test results, and imaging scans, in order to detect patterns and forecast a patient's medical state. This could assist physicians in formulating more precise diagnoses and devising treatment strategies that are more impactful.

The implementation of AI can aid in the enhancement of patient outcomes by predicting and preventing health problems proactively. By analyzing data from electronic health records and wearable devices, AI can find patients who are at risk of developing certain conditions, such as heart disease or diabetes. Healthcare professionals can then intervene early, with targeted interventions such as lifestyle changes or medication, to prevent or delay the onset of these conditions.

Another way AI is transforming healthcare is by automating administrative tasks and streamlining processes. It can help to reduce healthcare costs and free up healthcare professionals to focus on patient care. For example, AI-powered chatbots can help patients to schedule appointments, answer questions, and manage medication regimens, lowering down the need for administrative staff.

However, the integration of AI in healthcare also raises concerns around privacy, security, and ethics. It is important for healthcare professionals to be aware of these concerns and work to ensure that AI is used responsibly and in the best interests of patients.

5.4 AI and Ethics

The significance of AI and its ethical considerations is progressively gaining prominence. The use of AI in decision-making processes raises concerns about fairness, accountability, and transparency. It is necessary to establish ethical principles to guarantee the responsible and ethical development and utilization of AI.

One of the main ethical issues surrounding AI is bias. The level of bias in AI systems is directly linked to the bias present in the data they are trained on, and if the data is biased, it can result in discriminatory results. For example, facial recognition technology has been shown to be less accurate when identifying people of color, which can have serious consequences in law enforcement and other fields. It is essential to ensure that AI systems are trained on diverse data sets and that they are regularly audited for bias.

Another important ethical consideration is privacy. AI systems can collect and analyze vast amounts of personal data, which can be used for a variety of purposes. There is a need to ensure that individuals have control over their data and that it is used only for its intended purpose. To promote transparency, it is necessary to make sure that AI systems clearly disclose the data they gather and how they utilize it.

Accountability is another key ethical issue. As AI systems become more autonomous, it is important to ensure that there is a mechanism for holding them accountable for their decisions. It involves making sure that there is a well-defined hierarchy of accountability for any adverse consequences arising from the decisions made by AI.

In conclusion, AI has the potential to bring tremendous benefits to society, but it is important to ensure that it is developed and used in an ethical and responsible manner. This requires a collaborative effort from all stakeholders, including governments, businesses, and individuals. By working together, we can ensure that AI is used to improve our lives while respecting our rights and values.

5.5 AI and the Environment

AI will play a significant role in addressing environmental challenges such as climate change, resource depletion, and pollution. AI technologies have the potential to enhance resource utilization, track environmental factors, and create eco-friendly approaches.

A key method by which AI can contribute to environmental preservation is by streamlining energy consumption. By optimizing energy usage in structures and transport systems, AI can assist in lowering greenhouse gas emissions. Furthermore, AI's ability to manage electrical grids with greater efficiency can lessen the demand for additional power plants and transmission infrastructure.

AI can be employed to oversee and control natural resources as well. For instance, it can be utilized to track water quality and volume, identify contamination, and regulate fisheries. Additionally, AI can be applied to refine agricultural practices, minimizing the reliance on pesticides and fertilizers while enhancing crop production.

Another important use of AI is in developing sustainable solutions. AI can be harnessed to develop new materials and technologies that are more sustainable and environmentally friendly. As an illustration, AI can be harnessed to improve the design of wind turbines, solar panels, and electric vehicles, resulting in enhanced efficiency and cost-effectiveness.

By optimizing resource use, monitoring environmental conditions, and developing sustainable solutions, AI can help us build a more sustainable future. It is important to ensure that AI is developed and used in an ethical and responsible manner, taking into account the potential environmental impact of its use.

> ### *Did you know?*
>
> AI is improving disaster response: AI-powered drones and robots can be deployed in disaster-stricken areas to assess damage, locate survivors, and deliver aid more efficiently, ultimately saving lives and reducing the impact of natural disasters.

5.6 Bias and Fairness in AI

In the realm of AI, bias and fairness stand as two essential notions. Bias in AI refers to the systematic and unfair treatment of certain individuals or groups based on their race, gender, age, or other characteristics. Bias can occur in various stages of the AI development process, from data collection and model design to deployment and decision-making. For instance, biased data sets can lead to biased algorithms that perpetuate discriminatory practices, such as hiring or lending decisions. Biased models can also result from limited or skewed training data, flawed algorithms, or biased input from human developers.

On the other hand, fairness in AI refers to the equitable treatment of all individuals and groups, regardless of their background or identity. Fairness can be achieved through various strategies, such as designing unbiased algorithms, selecting diverse and representative data sets, and monitoring and auditing AI systems for bias. Fairness in AI is crucial because it can ensure that AI systems do not perpetuate social inequalities and biases, and that they uphold fundamental human rights and values.

To address bias and promote fairness in AI, researchers and practitioners have developed various approaches and techniques. Some of these techniques include data preprocessing and cleaning, which involves detecting and removing biased data points or attributes. Other techniques involve modifying the algorithm design to ensure that it does not produce biased outputs, or adding fairness constraints that ensure equitable treatment for all individuals and groups. Moreover, certain researchers champion increased transparency and responsibility in AI systems, for instance, via interpretable AI or ethical AI structures.

Despite these efforts, bias and fairness in AI remain significant challenges, as they often involve complex and systemic issues that require multidisciplinary approaches and collaboration. It is crucial for AI developers and stakeholders to understand the potential impact of AI on society and to prioritize ethical and human-centered design principles in AI development. By doing so, we can create AI systems that are not only accurate and efficient but also fair and equitable, and that benefit all members of society.

5.7 AI and Privacy

AI technologies collect and analyze vast amounts of personal data, including sensitive information such as health records, financial information, and biometric data. This raises concerns about data security and privacy protection.

A primary privacy issue associated with AI involves data breaches. Inadequately secured AI systems may be susceptible to hacking and various cyber-attacks. Consequently, sensitive information could be exposed, placing individuals in danger of identity theft and other fraudulent activities.

An additional worry is the possibility of data being exploited inappropriately. AI systems may be educated using data sets containing sensitive or private details, which could then be employed for purposes that individuals have not authorized. For instance, AI-driven advertisements can utilize personal information to target individuals with tailored ads, potentially resulting in undesired invasions of their private lives.

There is also a risk that AI systems could perpetuate or even exacerbate existing biases and discrimination. If AI systems are trained on biased data sets, they may produce biased or discriminatory outcomes. This can have serious implications for privacy and civil liberties, particularly in areas such as law enforcement and hiring practices.

In summary, AI possesses significant growth potential, but it is crucial to guarantee its responsible and ethical development and usage. This necessitates strong data security precautions, clear data handling procedures, and a dedication to preserving individuals' privacy rights.

5.8 Legal Liability

The rapid growth and increasing use of AI raise complex legal issues, including questions about legal liability. As AI systems become more autonomous and make decisions that have real-world consequences, it becomes important to consider who is responsible when things go wrong.

One aspect of legal liability for AI involves determining whether the creators, developers, or users of the technology should be held accountable for any harmful consequences that may arise from the system's actions. This is particularly relevant when AI systems cause physical or financial harm, infringe on privacy rights, or display biased behavior. As AI systems become more autonomous, the lines of responsibility may blur, complicating the process of assigning liability.

Additionally, as AI systems begin to interact with one another, determining fault in cases where multiple systems are involved becomes increasingly difficult. To address these challenges, some legal experts propose the establishment of new legal frameworks that recognize AI systems as legal entities with their own rights and responsibilities, similar to how corporations are treated under the law.

Furthermore, the dynamic nature of AI systems, including their ability to learn and adapt over time, raises questions about the appropriate time frame for assessing legal liability. Should developers be held responsible for unforeseen consequences that arise from a system's evolution over time, or is it more appropriate to focus on the users and operators who directly interact with the AI?

In conclusion, addressing the legal liability of AI will require a careful balance of interests, including the protection of individual rights, the promotion of innovation, and the need for accountability. As AI technology continues to evolve, it is essential for policymakers and legal experts to collaborate and develop appropriate regulations and frameworks that effectively address the unique challenges posed by AI.

CHAPTER 6
FUTURE OF AI

THE POWER OF AI

AI's future appears highly optimistic, poised to transform numerous sectors. The technology will evolve and become more refined, with machines improving their abilities to process, analyze, and decipher data. AI will substantially influence the labor market, automating numerous tasks presently performed by humans. Nevertheless, AI will also generate new employment prospects, particularly in fields like data science, machine learning, and robotics.

In the approaching future, AI will be increasingly pervasive in our everyday routines, as intelligent devices and systems become further ingrained in our homes and work environments. AI-driven virtual assistants and chatbots will proliferate, offering customized support and enhancing customer service interactions. Self-driving vehicles will gain a larger presence on our streets, decreasing accidents and boosting transportation effectiveness.

Over an extended period, the future of AI might encompass the creation of general AI, which has the ability to execute tasks that surpass its explicit programming. Such AI could result in substantial progress in areas like healthcare, finance, and education. Nonetheless, it also presents ethical and social challenges, including the possibility of job displacement, breaches of privacy, and independent decision-making.

Notwithstanding these obstacles, AI's future appears incredibly stimulating, with the possibilities for novel innovations and discoveries being nearly boundless. As the technology progresses and develops further, it is certain to revolutionize many facets of our existence, rendering our planet more efficient, interconnected, and astute.

6.1 Trends and Developments in AI

There are several trends and developments in AI that are shaping the future of this technology:

- Deep learning: This particular area of machine learning employs artificial neural networks to enable machines to learn and advance their functionality. Deep learning has found application in various domains, such as image and speech recognition, natural language processing, and robotics.

- Explainable AI: As AI becomes more sophisticated, it becomes harder to understand how it makes decisions. Explainable AI aims to make the decision-making process of AI systems transparent and interpretable, allowing humans to understand and trust the decisions made by these systems.

- Edge computing: Edge computing involves processing data on local devices, such as smartphones or IoT devices, instead of sending it to a centralized server. This approach reduces latency, improves performance, and enhances privacy, making it well-suited for AI applications.

- Autonomous systems: Self-driving cars and drones, among other autonomous systems, are becoming more prevalent. AI is employed in these systems to make judgments and operate independently of human intervention, increasing efficiency and minimizing the possibility of human mistakes.

- Humans and AI: With the increasing prominence of AI, it is crucial to establish effective collaboration between humans and machines. This collaboration necessitates the creation of interfaces and interaction techniques that facilitate seamless teamwork between humans and AI.

6.2 Opportunities and Challenges in AI

AI's development and usage present both possibilities and difficulties. Several crucial opportunities associated with AI include:

- Automation: AI-driven technologies provide the benefit of automating processes, which can alleviate the workload on human workers by taking over mundane and repetitive tasks. Machines are capable of performing these duties with greater efficiency and precision compared to humans, enabling people to focus on activities that demand higher levels of creativity and cognitive effort. This results in enhanced productivity, as AI-enabled systems can complete tasks more quickly and consistently. Moreover, automation contributes to a decrease in workplace incidents and injuries, since machines can manage tasks that pose risks or dangers to human workers.

- Efficiency: AI also presents a considerable opportunity in its ability to swiftly and accurately process vast quantities of data. Systems powered by artificial intelligence can examine intricate data sets more precisely and quickly than humans, enabling organizations to make well-informed, data-based decisions. This leads to heightened efficiency, as AI can detect patterns and trends in data that could be overlooked by human analysts. The extraordinary capacity of AI to handle data on a large scale and with remarkable precision allows organizations to fine-tune their operations, simplify processes, and conserve time and resources.

- Personalization: AI offers a substantial advantage in the realm of personalization, enabling organizations to adapt their products, services, and experiences to cater to the unique needs and tastes of each customer. Systems driven by artificial intelligence can assess customer information, including purchasing patterns and browsing habits, in order to produce customized suggestions, offers, and content. This results in a more individualized and bespoke experience for customers, enhancing their satisfaction and allegiance to the brand. Personalization also aids businesses in forging deeper connections with their clientele and acquiring a more profound comprehension of their desires and inclinations.

- Innovation: The ability of AI to foster novel advancements across various sectors is a key opportunity presented by this technology. In areas like healthcare, finance, and education, AI holds the potential to enable substantial progress and innovation. For example, AI can contribute to the early identification and diagnosis of illnesses, resulting in better treatment outcomes and more efficient therapeutic approaches. In the financial sector, AI can be employed to detect fraud, minimize risks, and enhance investment tactics. In education, AI can tailor learning experiences to individual students, promoting more personalized and effective teaching. Moreover, AI can be utilized in fields such as science and engineering to expedite research and development, paving the way for groundbreaking discoveries and innovations.

However, there are also several challenges associated with the development of AI, including:

- Bias: AI systems can be biased, leading to unfair or discriminatory outcomes, particularly if they are trained on biased data.

- Privacy: AI systems can collect and analyze vast amounts of personal data, raising concerns about privacy and security.

- Job displacement: As AI automates more tasks, it may lead to job displacement in certain industries, particularly those that rely on routine and repetitive tasks.

- Ethical considerations: There are several ethical considerations associated with AI, including concerns about the use of autonomous weapons and the potential for AI to make decisions that have significant ethical implications.

Achieving a balance between AI's opportunities and challenges is crucial to ensure its responsible and ethical development and deployment, resulting in benefits for society as a whole.

6.3 AI and the Singularity

AI and human augmentation are two distinct but related concepts that can be combined to enhance human capabilities and improve overall performance. Human augmentation involves enhancing human capabilities through technology, such as wearable devices or implants, while AI involves creating machines that can simulate human intelligence and perform tasks that would typically require human intervention.

The integration of AI and human augmentation has a significant advantage in that it enables us to surpass our cognitive restrictions. By swiftly and precisely analyzing vast amounts of data, AI can provide us with insights and recommendations that may not be apparent to humans. Meanwhile, human augmentation technologies can enhance our interaction with machines, improving our capacity to manipulate and control them. This way, we can capitalize on the strengths of both AI and humans, augmenting our intellectual abilities, and improving our problem-solving skills.

For example, a surgeon could use an AI-powered robotic arm to perform a complex surgery, with the AI system providing real-time feedback and analysis to guide the surgeon's movements. Similarly, a factory worker could wear an exoskeleton that enhances their strength and dexterity, while an AI system analyzes the production line data to identify areas for optimization and improvement.

However, combining AI and human augmentation also raises ethical and social concerns, particularly around issues of privacy and control. It is important to ensure that these technologies are developed and deployed in a responsible and ethical manner that respects individual rights and promotes social and economic equality.

In summary, AI and human augmentation possess the capacity to revolutionize numerous facets of our lives, enhancing performance, expanding our abilities, and improving our overall quality of life. Nevertheless, it is crucial to exercise caution and prudence when developing and employing these technologies, ensuring that they benefit society as a whole.

6.4 Quantum Computer and AI

Quantum computers perform computations using quantum mechanics and are distinct from classical computers that use binary digits (bits) to represent data. They leverage quantum bits (qubits) to perform computations, and as a result, have the capability to solve certain problems more efficiently than classical computers, including some problems that are pertinent to AI.

Quantum computers have the potential to enhance the performance of machine learning algorithms, which rely on mathematical optimization methods to identify the optimal parameters for a particular model. These optimization problems can be highly intricate and necessitate a significant number of computational resources. Quantum computers have the capacity to solve these optimization problems more rapidly than classical computers, enabling the training of more sophisticated models in a shorter duration of time. As a result, the use of quantum computers in AI can improve the overall performance of machine learning algorithms.

Another potential application of quantum computers in AI is to improve the accuracy of certain types of machine learning models. An instance of such a technology is the quantum support vector machines (QSVMs), which are a category of machine learning models utilized for classification tasks. QSVMs use a quantum algorithm to perform a specific step in the classification process, allowing them to achieve higher accuracy than classical support vector machines.

There are also some challenges associated with using quantum computers in AI. One challenge is that quantum computers are currently very expensive and difficult to build and operate. Another challenge is that not all problems in AI can be efficiently solved using quantum algorithms. Additionally, quantum computers are very sensitive to noise and errors, which can reduce their accuracy and make them less effective for certain types of computations.

Despite these challenges, researchers are actively exploring the potential of quantum computers in AI. Some companies, such as IBM and Google, have already developed early versions of quantum computers and are exploring their potential applications. As quantum computers continue to evolve and become more powerful, it is likely that they will play an increasingly important role in the development of AI.

> **_Did you know?_**
>
> Quantum computers operate at extremely low temperatures. To function effectively, quantum computers must be maintained at temperatures near absolute zero (approximately -273.15°C or -459.67°F). These ultra-cold conditions help minimize interference from the external environment, ensuring that the delicate quantum states are preserved.

6.4 AI and the Singularity

The concept of the Singularity refers to a hypothetical point in the future where artificial intelligence (AI) surpasses human intelligence, leading to a rapid and dramatic transformation of society. Some proponents of the Singularity suggest that this could result in a utopian future, while others warn of catastrophic consequences if we are not careful.

There is no consensus among AI experts about whether the Singularity will actually happen, and if so, when it might occur. Some argue that we are already experiencing the early stages of the Singularity, with AI systems becoming increasingly sophisticated and capable of performing tasks that were once the exclusive domain of humans.

However, many experts also caution that the Singularity is unlikely to occur in the near future, and that we should focus on more immediate concerns related to AI, such as ethical considerations and the impact of automation on employment.

Even if the Singularity does occur, it is unclear what the consequences would be. Some suggest that it could lead to a post-scarcity society where AI systems meet all of our needs, while others warn of the potential for an AI-dominated world where humans are marginalized or even destroyed.

The Singularity remains a highly speculative concept, and it is important to approach discussions of AI and its potential impact on society with caution and consideration. Rather than focusing on the possibility of a Singularity, it may be more productive to focus on the concrete ways in which AI is currently being used, and how we can ensure that it is developed and deployed in a responsible and ethical manner that benefits society as a whole.

CONCLUSION

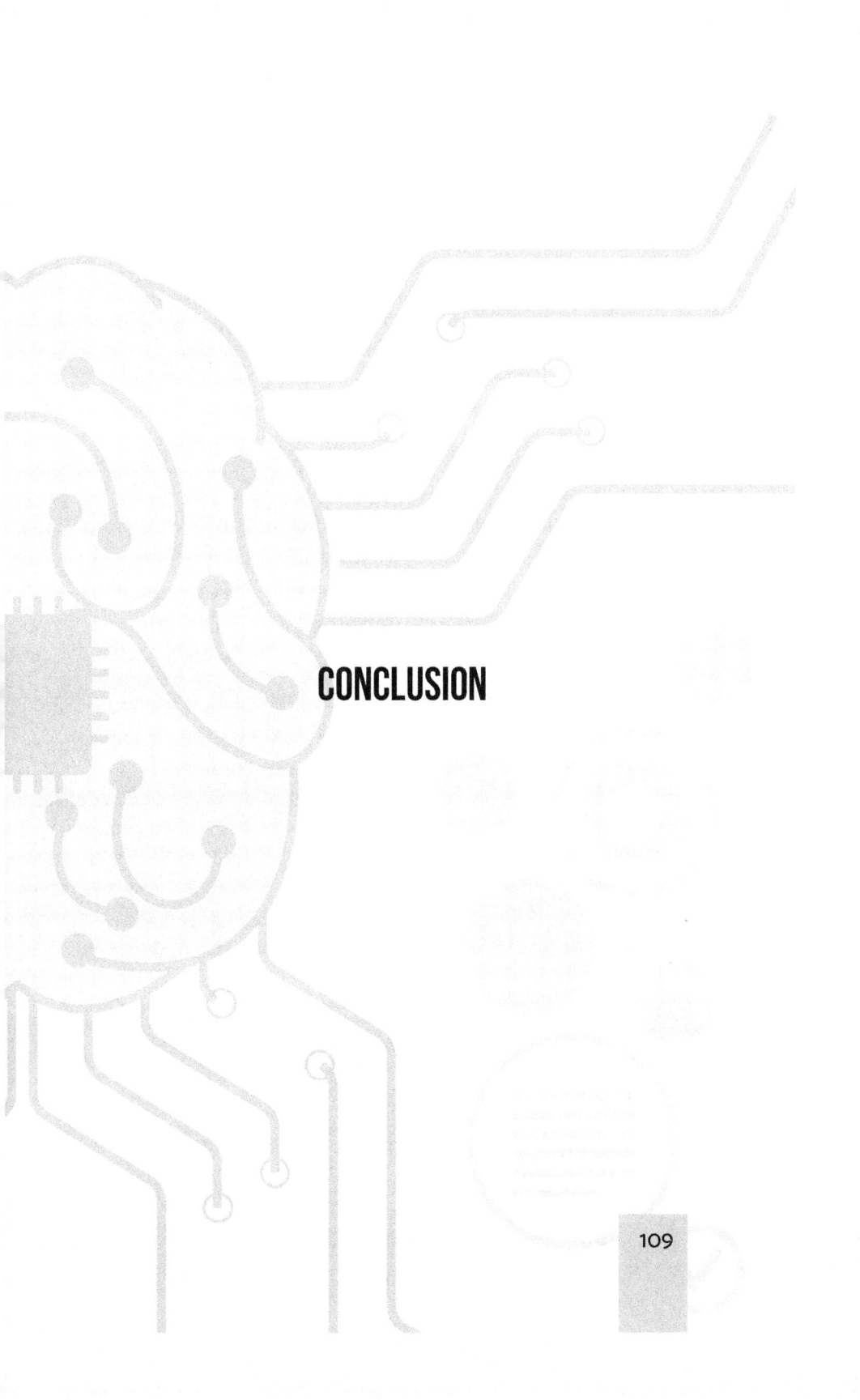

THE POWER OF AI

The future of AI presents a blend of exhilaration and hurdles. While AI holds the power to reshape various facets of our existence, encompassing healthcare, transportation, and leisure, it also demands that we address significant challenges to guarantee its ethical and responsible development and implementation.

One of AI's most prominent advantages is its capacity to automate monotonous tasks, thereby boosting productivity. This opens up opportunities for refining multiple sectors, such as customer service, manufacturing, and logistics. By streamlining operations, AI can potentially lower expenses and augment efficiency. Furthermore, AI's ability to scrutinize vast quantities of data, identify trends, and offer invaluable insights contributes to improved decision-making.

AI's potential to enhance healthcare is another notable benefit. AI-driven technologies can accurately diagnose illnesses, evaluate medical imagery, and devise tailored treatment strategies. These capabilities may lead to improved patient outcomes while simultaneously reducing overall healthcare costs.

However, numerous obstacles are associated with AI's development and deployment. One primary concern is the risk of bias and discrimination. AI systems may perpetuate or exacerbate existing biases and discriminatory practices, particularly when trained on inherently biased datasets. It is vital to ensure the equitable and unbiased development and implementation of AI systems.

Another challenge lies in AI's impact on the workforce. Automation of routine tasks may result in job displacement and a shift in sought-after skills. It is crucial to devise strategies that facilitate worker transitions to new roles and guarantee the fair distribution of AI's benefits.

Finally, the potential misuse of AI raises serious concerns, especially regarding cybersecurity and warfare. Since AI systems can be susceptible to cyber-attacks, it is essential to prioritize the responsible, ethical, and equitable development and application of AI. This approach requires a comprehensive evaluation of AI's risks and rewards, as well as a commitment to promoting the equitable distribution of its benefits. By doing so, we can harness AI's potential to foster a more prosperous future for all.

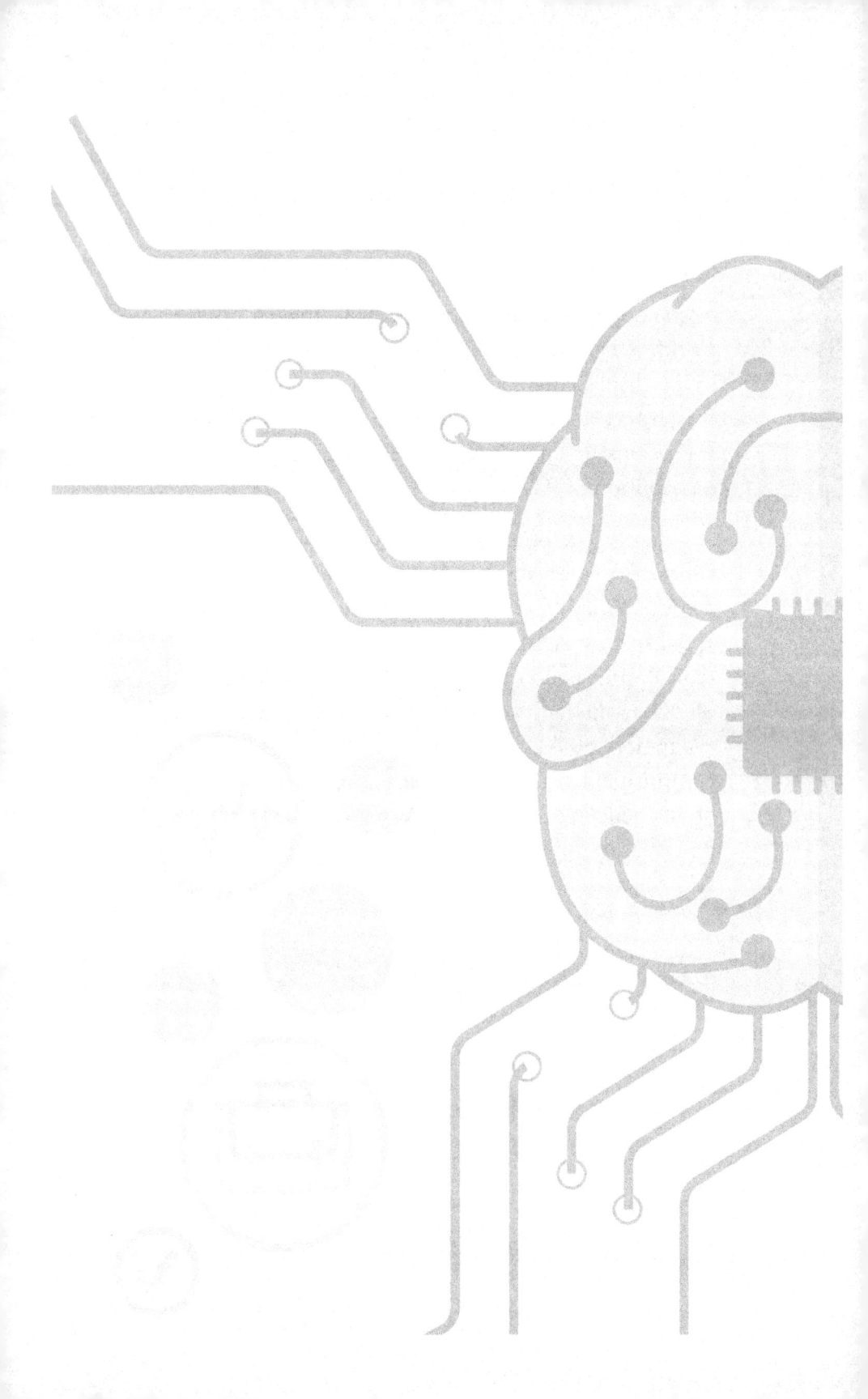

BIBLIOGRAPHY

1. Dennis, M. A. (2023, January 20). *Marvin Minsky | American scientist.* Encyclopedia Britannica. https://www.britannica.com/biography/Marvin-Lee-Minsky

2. *The brief history of artificial intelligence: The world has changed fast – what might be next?* (2022, December 6). Our World in Data. https://ourworldindata.org/brief-history-of-ai

3. McCarthy, J. J., Minsky, M., Rochester, N., & Shannon, C. E. (2006, December 15). *A Proposal for the Dartmouth Summer Research Project on Artificial Intelligence, August 31, 1955.* Ai Magazine; Association for the Advancement of Artificial Intelligence. https://doi.org/10.1609/aimag.v27i4.1904

4. *What is artificial intelligence in healthcare? | IBM.* (n.d.). https://www.ibm.com/topics/artificial-intelligence-healthcare#:~:text=AI%20and%20machine%20learning%20algorithms,race%2C%20ethnicity%20or%20income%20level.

5. *Artificial intelligence (AI) | Definition, Examples, Types, Applications, Companies, & Facts.* (2023, March 23). Encyclopedia Britannica. https://

www.britannica.com/technology/artificial-intelligence/Alan-Turing-and-the-beginning-of-AI

6. PricewaterhouseCoopers. (n.d.). *No longer science fiction, AI and robotics are transforming healthcare*. PwC. https://www.pwc.com/gx/en/industries/healthcare/publications/ai-robotics-new-health/transforming-healthcare.html

7. S. (2023, February 13). *AI in Manufacturing: Here's Everything You Should Know*. Simplilearn.com. https://www.simplilearn.com/growing-role-of-ai-in-manufacturing-industry-article#:~:text=Manufacturers%20use%20AI%20to%20analyse,to%20be%20scheduled%20in%20advance.

8. *What is AI Marketing? A Complete Guide | Marketing Evolution*. (n.d.). https://www.marketingevolution.com/marketing-essentials/ai-markeitng#:~:text=AI%20marketing%20can%20help%20you,re%2Dengage%20with%20the%20brand.

9. *Ethics of Artificial Intelligence*. (2023, February 3). UNESCO. https://www.unesco.org/en/artificial-intelligence/recommendation-ethics

10. Piluta, R. (2023, February 22). *Machine Learning in Healthcare: 12 Real-World Use Cases – NIX United.* NIX United – Custom Software Development Company in US. https://nix-united.com/blog/machine-learning-in-healthcare-12-real-world-use-cases-to-know/#:~:text=Machine%20learning%20%28ML%29%20is%20a%20subclass%20of%20artificial,instructed%20on%20exactly%20how%20to%20address%20the%20problem.

11. Rose, J. (2023, February 17). *Medical Imaging with AI: Revolutionizing the Healthcare Industry.* TheBlue.ai. https://theblue.ai/blog/trends-en/medical-imaging-blog/

12. Casella, D. (2022, September 9). *AI and privacy: Everything you need to know.* Telefonaktiebolaget LM Ericsson. https://www.ericsson.com/en/blog/2022/8/ai-and-privacy-everything-you-need-to-know#:~:text=For%20example%2C%20AI%20systems%20may,in%20question%20to%20unpredictable%20consequences.

13. Burns, E., & Brush, K. (2021, March

29). *deep learning.* Enterprise AI. https://www.techtarget.com/searchenterpriseai/definition/deep-learning-deep-neural-network#:~:text=Deep%20learning%20is%20a%20type,includes%20statistics%20and%20predictive%20modeling.

14. *What is Supervised Learning? | IBM.* (n.d.). https://www.ibm.com/topics/supervised-learning#:~:text=Supervised%20learning%2C%20also%20known%20as,data%20or%20predict%20outcomes%20accurately.

15. *What is Unsupervised Learning? | IBM.* (n.d.). https://www.ibm.com/topics/unsupervised-learning#:~:text=the%20next%20step-,What%20is%20unsupervised%20learning%3F,the%20need%20for%20human%20intervention.

16. Díaz, D. (2023, January 14). *What is reinforcement learning? - University of York.* University of York. https://online.york.ac.uk/what-is-reinforcement-learning/

17. *Machine Learning: 6 Real-World Examples.* (n.d.). Salesforce EMEA Blog. https://www.

salesforce.com/eu/blog/2020/06/real-world-examples-of-machine-learning.html

18. A., & Alkhaldi, N. (2022, September 8). *Predictive Analytics In Healthcare: 7 Examples and Risks — ITRex*. ITRex. https://itrexgroup.com/blog/predictive-analytics-in-healthcare-top-use-cases/

19. C. (2023a, February 9). *7 Machine Learning Algorithms to Know: A Beginner's Guide*. Coursera. https://www.coursera.org/articles/machine-learning-algorithms

20. Saxena, S. (2021, March 30). *Tokenization and Text Normalization*. Analytics Vidhya. https://www.analyticsvidhya.com/blog/2021/03/tokenization-and-text-normalization/#:~:text=Tokenization%20is%20the%20process%20of%20splitting%20a%20text,most%20common%20tokenization%20process%20is%20whitespace%2F%20unigram%20tokenization.

21. Husnain, A. (2023, March 23). *The Intersection of AI and Blockchain: Opportunities and Challenges*. TechBullion. https://techbullion.com/the-intersection-of-ai-and-blockchain-

opportunities-and-challenges/

22. Díaz, D. (2022, May 31). *The role of natural language processing in AI - University of York.* University of York. https://online.york.ac.uk/the-role-of-natural-language-processing-in-ai/

23. *Natural Language Processing (NLP): 7 Key Techniques.* (2021, October 19). MonkeyLearn Blog. https://monkeylearn.com/blog/natural-language-processing-techniques/

24. *What is Computer Vision? | IBM.* (n.d.). https://www.ibm.com/topics/computer-vision#:~:text=Resources-,What%20is%20computer%20vision%3F,recommendations%20based%20on%20that%20information.

25. Martin, A. (2022, November 1). *Robotics and artificial intelligence: The role of AI in robots.* AI Business. https://aibusiness.com/verticals/robotics-and-artificial-intelligence-the-role-of-ai-in-robots

26. Owen-Hill, A. (n.d.). *What's the Difference Between Robotics and Artificial Intelligence?* https://blog.robotiq.com/whats-the-difference-between-robotics-and-artificial-

intelligence

27. Manyika, J. M. (2022, October 10). *AI & Society*. American Academy of Arts & Sciences. https://www.amacad.org/daedalus/ai-society

28. *AI and Society: A vision for artificial intelligence research at University of Waterloo | Artificial Intelligence Group*. (n.d.). https://uwaterloo.ca/artificial-intelligence-group/ai-and-society/overview

29. Van Rijmenam Csp, M. (2022, October 12). *What is the Future of Artificial Intelligence?* Dr Mark Van Rijmenam, CSP | Strategic Futurist Keynote Speaker. https://www.thedigitalspeaker.com/future-artificial-intelligence/

30. Publications Office of the European Union. (2020, November 25). *Trends and developments in artificial intelligence : challenges to the intellectual property rights framework : final report*. &Copy; European Union. https://op.europa.eu/en/publication-detail/-/publication/394345a1-2ecf-11eb-b27b-01aa75ed71a1/language-en

31. Rosen, H. (2023, February 7). *Top Five*

Opportunities And Challenges Of AI In Healthcare. Forbes. https://www.forbes.com/sites/forbesbusinesscouncil/2023/02/07/top-five-opportunities-and-challenges-of-ai-in-healthcare/?sh=5dc9a2d22805

Printed in Great Britain
by Amazon